Sawsan Mosa

Use Multi Media Chemical Compounds for Treatment Wastewater

AF166522

Sawsan Mosa

Use Multi Media Chemical Compounds for Treatment Wastewater

LAP LAMBERT Academic Publishing

Impressum / Imprint
Bibliografische Information der Deutschen Nationalbibliothek: Die Deutsche Nationalbibliothek verzeichnet diese Publikation in der Deutschen Nationalbibliografie; detaillierte bibliografische Daten sind im Internet über http://dnb.d-nb.de abrufbar. Alle in diesem Buch genannten Marken und Produktnamen unterliegen warenzeichen-, marken- oder patentrechtlichem Schutz bzw. sind Warenzeichen oder eingetragene Warenzeichen der jeweiligen Inhaber. Die Wiedergabe von Marken, Produktnamen, Gebrauchsnamen, Handelsnamen, Warenbezeichnungen u.s.w. in diesem Werk berechtigt auch ohne besondere Kennzeichnung nicht zu der Annahme, dass solche Namen im Sinne der Warenzeichen- und Markenschutzgesetzgebung als frei zu betrachten wären und daher von jedermann benutzt werden dürften.

Bibliographic information published by the Deutsche Nationalbibliothek: The Deutsche Nationalbibliothek lists this publication in the Deutsche Nationalbibliografie; detailed bibliographic data are available in the Internet at http://dnb.d-nb.de.
Any brand names and product names mentioned in this book are subject to trademark, brand or patent protection and are trademarks or registered trademarks of their respective holders. The use of brand names, product names, common names, trade names, product descriptions etc. even without a particular marking in this works is in no way to be construed to mean that such names may be regarded as unrestricted in respect of trademark and brand protection legislation and could thus be used by anyone.

Coverbild / Cover image: www.ingimage.com

Verlag / Publisher:
LAP LAMBERT Academic Publishing
ist ein Imprint der / is a trademark of
OmniScriptum GmbH & Co. KG
Heinrich-Böcking-Str. 6-8, 66121 Saarbrücken, Deutschland / Germany
Email: info@lap-publishing.com

Herstellung: siehe letzte Seite /
Printed at: see last page
ISBN: 978-3-659-57880-9

The Table of Contents

page	Contents
1	INTRODUCTION
11	*Experimental*
12	*2.1- Photo and degradation experiments*
14	2.1.2 Experimental procedure
15	2.2- Removal of heavy metals
17	*Results*
18	3.1 -Photo catalytic reactor with sunlight source
35	Discussion
36	3.2- Decoloration of Murexide by hydrogen peroxide
42	3.3- Effect of different adsorbents on removal of heavy metals from waste water
49	Conclusion
51	References

I -INTRODUCTION

Survey explain the different methods which used to treatment of waste water

Quartz sand, which is called silica sand, is widely used in the construction of glass ceramic and foundry industry is a common non-metallic mineral materials. It also plays an important role in many new and high technological industries, such as semiconductor technology, SiO_2 thin-film material, atomic energy, optical fiber communication cable material and national defense science technology and others [(Niu et al., 2001,18; Tala, 2003, 40; Weige, 2001, 33; Wu et al., 2010, 1752)]. Activated carbon (AC) filter removes some harmful organic chemicals present in quantities above the Health Advisory Level (HAL). Included in these categories are pesticides, industrial solvents (halogenated hydrocarbons such as polychlorinated biphenyls PCB's) and polycyclic aromatic hydrocarbons (PAH's). THM's (Marawski A W, Kalenezuk R (2000). Adsorption of trihalomethanes (THMs) on to carbon spheres. Desalination; 130: 107-112). are a byproduct of the chlorination process that most public drinking water systems use for disinfection. Chloroform is the primary THM of concern. The EPA(examine in halation and dermal studies in developing the drinking water health goal for chlora amine) does not allow public systems to have more than 100 parts per billion (ppb) of THMs in their treated water. Some municipal systems have difficulty in meeting this standard activated carbon (AC) works by attracting and holding certain chemicals when water passes through it USEPA,"Stage 1 Disinfectants and Disinfection by – products rule", office of water (EPA). Viessman, (W. Hammer, M. Water supply and pollution control. Harper collies College Publishers. 15th edition. 1993: 485-488)]. Because of AC is a highly porous material having an extremely high surface area for contaminated adsorption. The carbon source is a variety of materials, such as peanut shells or coal. The raw carbon source is slowly heated in the absence of air to produce a high carbon material. The carbon is activated by passing oxidizing gases through the material

1

at extremely high temperatures. The activation process produces the pores that result in such high adsorptive properties. The adsorption process depends on many factors: Firstly, physical properties of the AC, such as pore size distribution and surface area; secondly, the chemical nature of the carbon source, or the amount of oxygen and hydrogen associated to it; thirdly, chemical composition and concentration of the contaminant; Fourth, temperature and pH of water; and fifth, the flow rate or time exposure of water to AC [(King W, Doddas L, Allen A (2000). *Envi Health Pers*; 108(9): 67-780)].

Physical properties

Forces of physical attraction or adsorption of contaminants to the pore walls is the most important AC filtration process. The amount and distribution of pores play key roles in determining how well contaminants are filtered. The best filtration occurs when pores are barely large enough to admit the contaminant molecule. Because contaminants come in many different sizes, they are attracted differently depending on pore size of the filter. In general, AC filters are most effective in removing contaminants that have relatively large molecules most organic chemicals[(Aziz, H.A.; Foul, A.A.; Isa, M.H.; Hung, Y-T. Physico-chemical. Int. J. Environ. Waste Manage. **2010**, 5, 269-285)]. The type of raw carbon material and its method of activation will affect the types of contaminants that are adsorbed. This is largely due to the influence that raw material and activation have on pore size and distribution.

Chemical properties

Processes other than physical attraction also affect AC filtration. The filter surface may actually interact chemically with organic molecules. Also, electrical forces between the AC surface and some contaminants may result in adsorption or ion exchange. Adsorption, then, is also affected by the chemical nature of the adsorbing surface. The chemical properties of the adsorbing surface are determined to a large extent by the activation process. AC materials of different activation processes will have chemical

2

properties that make them more or less attractive to various contaminants. For example, chloroform is better adsorbed by AC [(Wietlik J, Stanis R, Bodzek M (2002). *Wat res*; 36(9): 2328-36)] having a least amount of oxygen associated with the pore surfaces. The consumer can't possibly determine the chemical nature of an AC filter. Different types of AC filters have varying levels of effectiveness in treating different chemicals. The manufacturer should be consulted to determine if their filter will adequately treat the consumer's specific water problem.

Contaminant properties

Large organic molecules are the most effectively adsorbed by AC. A general rule of thumb is, similar materials tend to associate. Organic molecules and activated carbon are similar materials; therefore there is a stronger tendency for most of organic chemicals to associate with the activated carbon in the filter rather than staying dissolved in a dissimilar material like water. Generally, the least soluble organic molecules are most strongly adsorbed. Often the smaller organic molecules are held tightly because they fit into the smaller pores. The concentration of organic contaminants can affect the adsorption process. One AC filter may be more effective than another type at low contaminant concentrations, but may be less effective than the other filter at high concentrations. This type of behavior has been observed with chloroform removal [Wietlik J, Stanis R, Bodzek M (2002). *Wat res*; 36(9): 2328-36)]. The filter manufacturer should be consulted to determine how the filter will perform for specific chemicals at different levels of contamination.

According to the former researches [Dye wastewater with deep chromaticity and high chemical oxygen demand (COD), is one of industrial wastewater [(Wietlik J, Stanis R, Bodzek M (2002). *Wat res*; 36(9): 2328-36)] which is difficult to deal. Thus, the elementary task is decoloration. However, owing to the second pollution caused by the physicochemical and the biochemical methods, we should not abuse it. Actually the multistage series-connection dealing method of flocculent condensed precipitation-

3

biological oxidation cannot meet the emission standard. Therefore, it seems a wrathful issue to try to put forward a economical, efficient and effective treatment. The experiment investigated decoloration rate of the mixed solution under visible light, which was consisted of hydrogen peroxide and the dye solution. Then, our research work had focused on influence factors. The successful conclusion will provide a reference for making use of waste circularly for environment comprehensive management.

Azo dyes are a well-known class of dyes that are of high toxicity and even carcinogenic to the animals and human and they are not readily degradable. They are characterized by azo bonds ($-N=N-$) chromophores. The traditional physical, chemical and biologic means of wastewater treatment often have little degradation effect on this kind of pollutants. On the contrary, the technology of nano particulate TiO_2 photodegradation has been proved to be effective to them. Compared with the other conventional wastewater treatment means, this technology has such advantages as (A.P. Pantelis, P.X. Nikolaos, M. Dionissios,Water Res. 40 (2006)) wide application, especially to the molecule-structure complexed contaminants which cannot be easily degraded by the traditionalmethods; (J. Ara˜na, J.A. Herrera Melián) the TiO_2 itself has no toxicity to the health of our human livings and (H.D. Mansilla, C. Bravo, A: Chem. 181 (2006) 188) , it demonstrates a strong destructive power to the pollutants and can mineralize the pollutants into CO_2 and H_2O. Due to the excellent features of this technology, it appears promising and has drawn the attention of researchers of at home and abroad. Azo dyes represent a major class of synthetic organic pigments that are manufactured worldwide and have a variety of applications such as textiles, paper, foodstuff, and cosmetic. The toxicity and carcinogenic nature of these dyes and their precursors pose a threat to the environment Moreover; their degradation often leads to the formation of highly carcinogenic aromatic amines. For example p-amino azo

4

benzene has been classified as a carcinogenic compound and there has been a restriction on the production of dyes based on this molecule (Rajeshwar and Ibanez, 1997). However azo dyes continue to be a source of pollution in industrial processes, which utilize dyes to colour paper, plastics as well as natural and artificial fibers. Wastewaters from dyeing industries are released into nearby land or rivers without any treatment because the conventional treatment methods are not cost effective. In recent years, photo catalytic degradation has attracted increasing attention as cleaner and greener technology for removal of toxic organic and inorganic pollutants in water and wastewater (Parsons, 2004). Semiconductor photo catalysis appears to be a promising technology that has a number of applications in environmental system such as air purification, water disinfection, water purification, and hazardous waste remediation. Hydrogen peroxide photochemical degradation of organic pollutants in general, and a dye in particular in wastewater is a favored and promising technique (Egerton, 1997). The organics are completely mineralized into water and CO_2 without generating any harmful byproducts. This technique has been employed for the photomineralisation of large number of dyes such as methylene blue.

The pollution problems due to textile industry effluents have increased in the last years. The lost dye in the effluents reaches about 50% because the dying processes have a low yield [Pierce J., Dyers Color 110 (1994) 131]. Azo-dyes represent 60–70% of the 10,000 commercial dyes currently in use [Neamtu M., Siminiceanu I., Dyes Pigm. 53 (2002) 93),) (Sun J. H., Sun S. P., Dyes Pigm. 74 (2007) 647) , (Méndez-Paz D., Omil F. and Water Res. 39 (2005) 771) and (Brown M. A. and DeVito S. C., and Technology 23 (1993) 249). These dyes affect the environment and workers in textile industry. It was observed that the rate of bladder cancer in workers in the field of dye manufacturing increased (Selvam P. P., and Sivanesan S., J. Hazard. Mater. 155 (2008)39). Mordant dyes are one class of azo-dyes. Synthetic mordant dyes, representing about 30% of dyes used for wool, are especially useful for black and navy

5

shades. It is very important to remove these dyes from industrial effluents. Different techniques have been used to solve the problems caused by the toxic substances contained in industrial effluents including dyes such as membrane filtration, adsorption (Jae-Wook L., and Hee M., Dyes and Pigments, 69 (2006) 196).

Advanced oxidation processes (AOPs) are the processes which involve the accelerated production of the hydroxyl free radical, which is very reactive. Hydrogen peroxide alone produces hydroxyl radicals when it is added to water or wastewater; however, the rate of decolorization by H_2O_2 is extremely slow for many dyes [Pagga U. and Drown D., Chemosphere 15 (1986) 479].

The Fenton reaction discovered by Fenton H. J. in 1894 [Chen J. and Zhu L., *with* Catal. Today. 126 (2007) 463). Forty years later the Haber-Weiss mechanism was postulated, which revealed that the effective oxidative agent in the Fenton reaction was the hydroxyl radical (Lucking F., Koser H., and Jank M., *aqueous solution*", Water Res. 32 (1998) 2607). Fenton reaction process has tight working pH range (pH 2–4). In Fenton reaction hydroxyl radicals are generated as shown in equation 1.

$$Fe^{2+} + H_2O_2 \longrightarrow Fe^{3+} + HO^{\cdot} + HO^{-} \qquad (1)$$

In recent years much attention has been given on the treatment of azo dyes by zero-valent iron (ZVI) (Chang M.C., Shu H.Y., Yu H.H., Sung Y.C., J. Chem. Technol. Biotechnol., 81 (2006) 1259) The advantages of the ZVI decolorization process include the ease in use as a pre-treatment process, easy recycling of the spent iron powder by magnetism as well as low iron concentration remaining and no necessity for further treatment of the effluents.

.

More than 10,000 dyes have been widely used in textile, paper, rubber, plastics, leather and cosmetic, pharmaceutical, and food industries. The discharge of colored wastes into the receiving water bodies not only affects their aesthetic nature but also interferes with the transmission of sunlight and therefore reduces the photosynthetic activity [Sawsan Mohamed Abu El Hassan Mosa1Life Science Journal 2014;11(2)]. As dyes are designed to resist breakdown with time, exposure to sunlight, water, soap, and oxidizing agent cannot be easily removed by conventional waste water treatment processes due to their complex structure and synthetic origins. Methylene blue (MB) is cationic dyes. MB will cause increased heart rate, vomiting, shock, Heinz body formation, cyanosis, jaundice, quadriplegia, and tissue necrosis in humans. Various conventional methods such as physical, chemical, and biological processes have been tried for the removal of dyes from aquatic media. Adsorption is one of the physical-chemical methods, which is found to be the most simple and economical to remove the dyes from effluents. The adsorption attempts have been made to find alternative low-cost adsorbents. Activated carbon prepared from these wastes helps to solve the waste disposal problem. Most of the activated carbons are produced by a two-stage process carbonization followed by activation. The first step is to enrich the carbon content and to create an initial porosity and the activation process helps in enhancing the pore structure. Basically, the activations are two different processes for the preparation of activated carbon: physical activation and chemical activation. Among the numerous dehydrating agents, zinc chloride in particular is the widely used chemical agent in the preparation of activated carbon. Knowledge of different variables during the activation process is very important in developing the porosity of carbon sought for a given application. Chemical activation by zinc chloride improves the pore development in the carbon structure, and because of the effect of chemicals, the yields of carbon are usually high.

Great variety of organic contaminants in a wide range of concentrations. many of these compounds are hard to treat by conventional activated sludge systems and may be finally released to the environment. Coagulation, flocculation, active carbon adsorption or membrane techniques can only transfer, more or less effectively, the contaminants from one phase to another, leaving the final environmental problem unsolved. Therefore it is more and more necessary to develop destructive systems leading to complete mineralization or at least to less harmful or easy-to- treat compounds and use gamma ray in treatment is much cost and danger from this point of view, oxidation of organic pollutants is an attractive method. However, the direct reaction of organic compounds with oxygen requires very high temperatures and involves high costs. During the last decade a series of new methods for water and waste water purification. The so called advanced oxidation processes (AoPs) (I.A. Alaton, I.A. Balcioglu and D.W. Bahnemann .H_2O_2/UV-C and TiO_2/UV-A processes. J. Water res. 36: 1143-1154, (2002)), (M. Muruganandham and M. Swaminathan Dyes and Pigments 62 269.(2007) ,((S.K. Chaudhur and B. Sur J. Environ. Engg. 126: 583,(2000)) and ,(D. Georgiou, P.Melidis and A. Aivasidis. J. Dyes and Pigments 52: 69,(2002)), have received considerable attention [Y. Yang, D.T.Wyatt and M. Bahorsky Decolonization of dyes using UV/H_2O_2 photochemical oxidation. J. Textile Chem. Colorist 30: 27,(1998).B. Neppolian, H.C. Choi and S. Sakthivel V. Murugesan, Solar/UV- induced photocatalytic degradation of three commercial textile dyes. J. Hazard. Mater. B 89: 303,(2002).Y. Wan J. Water res. 34 : 990,(2000).],[M. Neamtu, I. Siminiceanu and A. Yediler Kinetics aqueous J. Dyes and Pigments, 53: 93-99, (2002)]and[H.Nilsum J. Dyes Pigments 41:225-230,(1999)].They have been defined as near ambient temperature processes involving the generation of short- lived, highly oxidative species, especially the hydroxyl radical (redox potential = 2.8v). In principle, AOPs are characterized by high oxidation rates flexibility, small dimension of the equipment, and easy adaptability to water recycling processes. In comparison with other AOPs such as fenton, ozone, UV/O_3, etc., the photolysis of hydrogen peroxide shows some advantages such as the complete miscibility of $H_2O_2$with water, the stability and commercial

8

Availability of hydrogen peroxide, no phase transfer problems, no sludge formation simplicity of operation and laws investment costs photolysis of hydrogen peroxide is a very simple reaction running with high efficiency. In the present paper we have selected the UV photolysis of H_2O_2 [S.Yana, Wangp, yangx, Shanl, and Zhanndgw J. Hazard Mater , 179: 1-3, (2010)]and [G. M. Colonna and T. Caronna. J. Dyes and Pigments 41, 211-220,(1999)]. As a source of hydroxyl radicals and investigated the decolonization and mineralization of dye as a first step of a research amide at evaluating the applicability of this technique to the applicability of this technique to the treatment and re-use of dye house waste water.

The problems of the ecosystem are increasing with developing technology. Heavy metal and dyes pollution is one of the main problems. Toxic metal compounds coming to the earth's surface not only reach the earth's waters (seas, lakes, ponds and reservoirs), but can also contaminate underground water in trace amounts by leaking from the soil after rain and snow. Therefore, the earth's waters may contain various toxic metals. Drinking water is obtained from springs which may be contaminated by various toxic metals. One of the most important problems is the accumulation of toxic metals in food structures. As a result of accumulation, the concentrations of metals can be more than those in water and air. The contaminated food can cause poisoning in humans and animals. Although some heavy metals are necessary for the growth of plants, after certain concentrations heavy metals become poisonous for both plants and heavy metal microorganisms. Another important risk on corning contamination is the accumulation of these substances in the soil in the long term. It has been determined that various metal ions hinder various enzymes responsible for mineralization of organic compounds in the earth. Therefore, studies on the removal of heavy metal pollution are increasing. The purpose of this study was to investigate the removal of some toxic heavy metals from aqueous solution by adsorption, to determine the optimum removal

9

condition by using different types of adsorbents. Activated carbon is widely used as an adsorbent in industry due to its high adsorption capacity. This capacity is related to the pore b structure and chemical nature of the carbon surface in connection with preparation conditions. The "Ceramic Membrane Filtration System" is a reliable technology to produce clean water by removing the turbidity, bacteria, and cryptosporidium and other protozoa contained in raw water sources such as surface water and ground water. Using a unique ceramic membrane as a filter, this system is a low cost and long life filtration system. Therefore it can enable a water supply system to meet recent demands for safe and tasty water. In recent years; there is an increase interest in using non-chemical and low-cost adsorbent to remove heavy metals from wastewater. Mesoporous silica materials have attracted attention because of their utilities in adsorption, selective separation and catalysis MCM-41, one of the important mesoporous materials, has excellent periodicities in the mesoporous channels, larger BET surface area, high porosities and narrow pore sizes. Initial studies of the self assembled silica having a two-imensional hexagonal ordering of cylindrical mesopores stimulated activities on the preparation of several mesoporous materials using alkyl trimethyl ammonium surfactants of varying alkyl chain length as structure directing agents. Metal ion adsorbents have been prepared by grafting thiol functional groups as a monolayer onto the inner surface of MCM-41.Thiol functionalized MCM-41 was proved to be efficient adsorbents for mercury and heavy metal ions. A few other reports are available on synthesis thiol and amine functionalized mesoporous materials and use of the functionalized MCM-41 as adsorbents for removal of heavy metal ions [Sawsan Mohamed Abu El Hassan MOSA. Science Journal of Chemistry 2014; 2(1): 1-5]

II- Experimental

2.1- Photo and degradation experiments

2.1.1-Reagents

Analytical grade 30% H_2O_2, Methyl orange (MO), Eriochrom black T (EBT), 2,6-Dichlorophenol-indophenol sodium salt dehydrate (DID), and Xylenol orange (XO) are obtained from Sigma Chem. GmbH (Germany) and Ingredients water color E 102-133 is obtained from (Malta). Quartz, sand, Murexide, hydrogen peroxide, and activated carbon all compounds are analytical grad.

Methyl orange (MeO) in acidic medium

Methyl orange (MeO) in basic medium

Eriochrom black T

Xylenol orange (XO)

13

Structure of 2,6- Dichlorophenol-indophenol sodium salt dehydrate

Structure of Murexide

2.1.2 Experimental procedure

1) Preparation of silica sand

Weight 5g silica sand, and heat them for 10 min. at 50 C°

2) Decoloration of Murexide

i-   Preparation of 0.005M Murexide solution

Add 5 ml of H_2O_2 (30%) to 20 ml of Murexide solution and measure the absorbance of upper clear solution at 530 nm wavelength.

ii- Measure the absorbance changes of solution after left it in the visible light for 1, 2, 3, 4, 5, 6, 22, and 36 hr.

iii-Preparation of five solutions of 0.005 M Murexide, then stir each of them at different time at 25°C.

14

Take 20ml of this solution and add 5 g of quartz sand then measure the absorbance of upper clear solution at 530nm wavelength. Measure the changes of solution after left them in visible light for 1, 2, 3, 4 hr.

iv- Preparation of five solutions of 0.005 M Murexide then stir them at different time . To 20mL of these solutions, 5 g of quartz sand and 5g of activated carbon were added then, the absorbance of upper clear solution was measured at 530nm wavelength. Measure the changes of the absorbance of solutions after left them in visible light for 5, 10, 15, 30 min.

2.2 Removal of heavy metals
2.2.1 Experimental

All Chemicals Analytical grad: EDTA, $ZnSO_4$, $Mg(NO_3)_2$, $Ca(NO_3)_2$ $Hg(NO_3)_2$, $Pb(NO_3)_2$, NH_4Cl, NH_4OH, Murexide, and EBT. Adsorbents: activated carbon, ceramic powder and silica. In this study, the adsorption of Zn^{2+}, Mg^{2+}, Ca^{2+}, Pb^{2+} and Hg^{2+} on commercial activated carbon, silica, and ceramic were investigated. One gram of adsorbents with 5 ml of initial concentrations (C^o) of 0.01 Molar of heavy metals was shaken at 15 mints. Amounts of sample were taken from solutions containing adsorbent. The solution was diluted and titration with EDTA Solution. The equilibrium concentrations of heavy metals were determined after taking amount of samples from clear parts of solutions containing adsorbent and doing proper dilutions. The amount adsorbed (C^o) was calculated from the difference between initial and equilibrium concentrations. Which adsorbent does more adsorption at 24 hours was determined without looking at the equilibrium contact times. For this, 1g of adsorbents with 5 ml of initial concentrations of0.01 Molar of these ions for determination of the percent of removal by adsorbents of heavy metals were shaken separately for 15 mints at 30C°. At

15

the end of this period, the residual concentrations of heavy metals which were not adsorbed were determined with titration of EDTA.

1- Preparation of 0.01 M of ions solutions . In this study, the adsorption of Mg^{2+}, Ca^{2+}, Hg^{2+}, Pb^{2+} and Zn^{+2} on 2514 commercial activated carbon , ceramic and silica were investigated.

2- One gram of each absorbent with 5 ml of initial concentrations of ions and string them 15 min.

3- Preparation of 0.01 M of EDTA solution.

4- Standardization of EDTA by $ZnSO_4$.

5- Determination initial concentration of different ions by use EDTA before treatment.

6- Determination finial concentration of different ions by use EDTA after treatment.

7 - Titration each solution of ions before and after treatment by EDTA.

8- The percentage removal of elements and the amount of element adsorbed on adsorbent (qe) were:

calculated, respectively, as follows:

$\%Removal = (C_0-Ce/C_0) \times 100$

$qe = (C_0-Ce/M)V$

where qe is the amount of dye adsorbed on adsorbent at equilibrium, C_0 and Ce are the initial and equilibrium of element concentration in solution, respectively, V is the volume of solution (L), and M is the weight of adsorbent (g).

III- Results

3.1 Photo catalytic reactor with sunlight source:

Outdoor experiment was carried out with 100 ml Pyrex glass reservoir placed outside the laboratory building. Pyrex glass beaker containing (50 ml) and (0.5 ml) of 30% H_2O_2 [C. Jeed, and A. Mahata]. The degraded solution was taken for spectra measurement at various times.

Absorbance measurement: Four concentrations for every dyes (10^{-4}, 10^{-5}, 10^{-6} and 10^{-7} M) were used to find the effect of initial dyes concentrations on color removal. Also all of these experiments were performed in different time intervals. "absence of illumination" 1h,2h,3h,4h,6h,8h,10h,12h,week, 2- weeks, and 4-weeks). Temporal changes in the concentration of dyes were monitored by examining the variation in the maximal absorption. The absorbance's of dyes solutions before and after degradation was measured at different degradation times. The percentage of degradation was calculated from the following equation:

$$Degradation\% = [1-At/A_0] \times 100 \quad [\text{M.N Rashed and A.A. El-Amin}]$$

3 –Results

1-Photodegradation of MO in H_2O_2 is the best at sunlight irradiation for high concreted 3×10^{-7} M after 4h show table (3.1.3).

2-Photodegradation of EBT in H_2O_2 is the best at sunlight irradiation for high conctrated 3×10^{-5} M after 15h show table (3.1.7).

3-Photodegradation of XO in H_2O_2 is the best at sunlight irradiation for high conctrated 3×10^{-5} M after 4h show table(3.1.3).

4-Photodegradation of DID in H_2O_2 is the best at sunlight irradiation for high concentrated 6.1×10^{-5} M after 10h show table(3.1.5).

5-Photodegradation of dye food in H_2O_2 is the best at sunlight irradiation for high concentrated after 4weeks show table (3.1.9).

Schematic diagram are illustrated in (Figures (3.1.1), (3.1.2), (3.1.3) ,(3.1.4), 3.1.5)

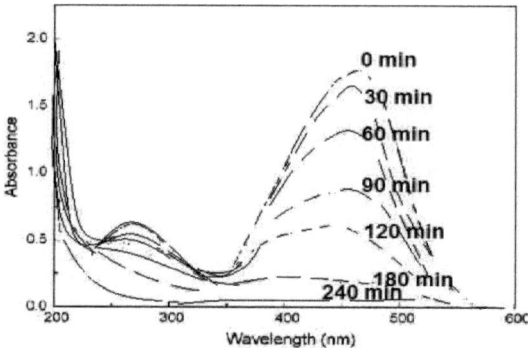

Fig(3.1.1). Effect of sunlight on MO at different time and [H₂O₂] = 0.1%.

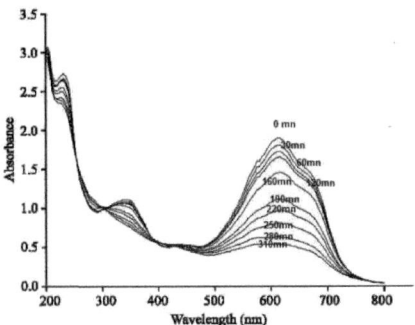

Fig (3.1.2). Effect of sunlight on EBT at different time and [H₂O₂] = 0.1%

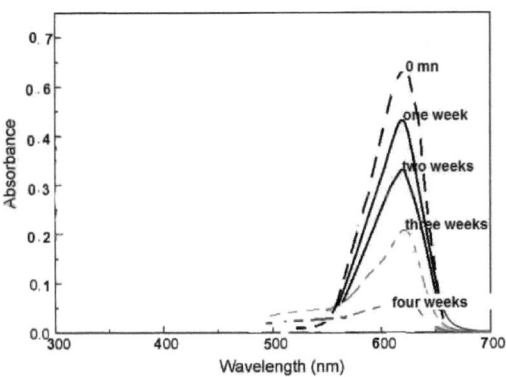

Fig(3.1.3). Effect of sunlight on Intgredints water color   E102-133 (commercial dye) at different time and [H₂O₂] =1%

20

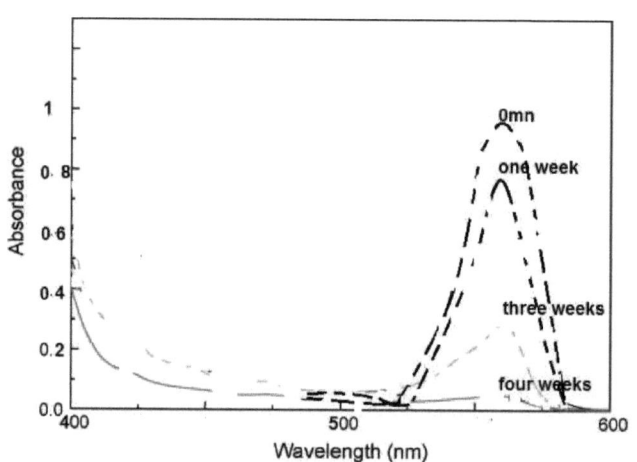

Fig(3.1.4). Effect of sunlight on DID at different time and [H₂O₂] =1%

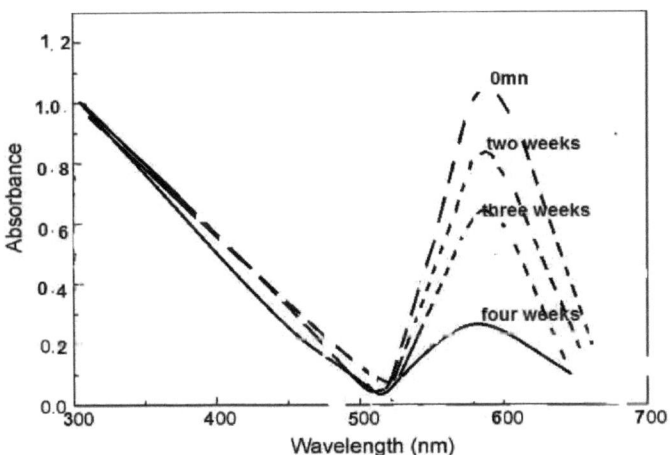

Fig(3.1.5). Effect of sunlight on XO at different time and [H₂O₂] =1%

Fig(3.1.5). Effect of sunlight on XO at different time and [H_2O_2] =1%

Table (3.1.1) **Effect of sunlight on different organic compounds in the presence of 0.1 % H$_2$O$_2$ and time of degradation is one hour.**

Dyes	Initial conc.	Initial ABC	Final ABC	Percentage of degradation	pH before	pH after
2,6-dichlorophenol-indophenol	6.1×10^{-5}	0.82	0.48	58.5%	7.8	7.3
Eriochrome black T	3×10^{-5}	1.8	1.4	22.22%	7.25	6.9
Methyl orange	3×10^{-7}	1.8	0.1	94%	8	7.9
commercial dye	1%	0.65	0.65	0%	7	7
Xylenol orange	3×10^{-5}	1.01	0.55	45.5%	10	9.5

Table (3.1.2) **Effect of sunlight on different organic compounds in the presence of 0.1 % H_2O_2 and time of degradation is 3 hour**

Dyes	Initial conc.	initial ABC	Final ABC	Percentage of degradation	pH before	pH after
2,6-dichlorophenol-indophenol	6.1×10^{-5}	0. 82	0.42	48%	7.8	6.9
Eriochrome black T	3×10^{-5}	1.8	1	44.44%	7.25	6.6
Methyl orange	3×10^{-7}	1.8	0.08	95%	8	7.8
Commercial dye	1%	0.65	0.65	0%	7	7
Xylenol orange	3×10^{-5}	1.01	0.40	60%	10	9.5

Table (3.1.3) Effect of sunlight on different organic compounds in the presence
0 .1 % H_2O_2 and time of degradation is 4 hour.

Dyes	Initial conc.	initial ABC	final ABC	Percentage of degradation	pH before	pH after
2,6-dichlorophenol-indophenol	$6.1×10^{-5}$	0.82	0.61	25%	7.8	6.8
Eriochrome black T	$3×10^{-5}$	1.8	0.8	55.55%	7.25	6.9
Methyl orange	$3×10^{-7}$	1.8	0.06	66.66%	8	7.5
Commercial dye	1%	0.65	0.65	0%	7	7
Xylenol orange	$3×10^{-5}$	1.01	0.35	65%	10	9.9

25

Table (3.1.4) Effect of sunlight on different organic compounds in the presence 0 .1 % H$_2$O$_2$ and time of degradation is 7 hour.

Dyes	Initial conc. M	initial ABC	final ABC	Percentage of degradation	pH before	pH after
2,6-dichlorophenol-indophenol	6.1×10^{-5}	0.82	0.40	51%	7.8	6.8
Eriochrome black T	3×10^{-5}	1.8	1	44.44%	7.25	6.9
Methyl orange	3×10^{-7}	1.8	0.01	100%	8	7.5
Commercial dye	1%	0.65	0.65	0%	7	7
Xylenol orange	3×10^{-5}	1.01	0.30	70%	10	9.8

Table (3.1.5)   Effect of   sunlight on different organic compounds in the presence 0 .1 % H_2O_2 and time of degradation is 9 hour.

Dyes	Initial conc. M	initial ABC	Final ABC	Percentage of degradation	pH before	pH after
2,6-dichlorophenol-indophenol	6.1×10^{-5}	0.82	0.01	100%	7.8	6.3
Eriochrome black T	3×10^{-5}	1.8	0.2	88.88%	7.25	6.7
Methyl orange	3×10^{-7}	1.8	0.01	100%	8	7.5
Commercial dye	1%	0.65	0.65	0%	7	7
Xylenol orange	3×10^{-5}	1.01	0.25	75%	10	9.5

Table (3.1.6) Effect of sunlight on different organic compounds in the presence 0 .1 % H_2O_2 and time of degradation is 10 hour.

Dyes	Initial conc. M	initial ABC	Final ABC	Percentage of degradation	pH buffer	pH after
2,6-dichlorophenol-indophenol	6.1×10^{-4}	0.82	0.01	100%	7.8	6.3
Eriochrome black T	3×10^{-5}	1.8	0.15	91%	7 .25	6.5
Methyl orange	3×10^{-7}	1.8	0.01	100%	8	7.4
Commercial dye	1%	0.65	0.65	0%	7	7
Xylenol orange	3×10^{-5}	1.01	0.20	80%	10	9.4

28

Table (3.1.7)   Effect of   sunlight on different organic compounds in the   presence 0 .1 % H_2O_2 and time of degradation is 15 hour.

Dyes	Initial conc.	initial ABC	Final ABC	Percentage of degradation	pH before	pH after
2,6-dichlorophenol-indophenol	6.1×10^{-5}	0.82	0.01	100%	7.8	6.3
Eriochrome black T	3×10^{-5}	1.8	0.01	100%	7.25	6.5
Methyl orange	3×10^{-7}	1.8	0.01	100%	8	7.4
Xylenol orange	3×10^{-5}	1.01	0.01	100%	10	9.4
Commercial dye	1%	0.65	0.65	0%	7	7

Table (3.1.8) Effect of sunlight on different organic compounds in the presence 0 .1 % H_2O_2 and time of degradation is one week.

Dyes	Initial conc.	Initial ABC	Final ABC	Percentage of degradation	pH before	pH after
2,6-dichlorophenol-indophenol	6.1×10^{-5}	0.82	0.01	100%	7.8	7.7
Eriochrome black T	3×10^{-5}	1.8	0.01	100%	7.25	7.1
Methyl orange	3×10^{-7}	1.8	0.0	100%	8	7.4
Xylenol orange	3×10^{-5}	1.01	0.01	100%	10	9.4
Commercial dye	1%	0.65	0.35	46%	7	6.9

30

Table (3.1.9) Effect of sunlight on different organic compounds in the presence 0 .1 % H$_2$O$_2$ and time of degradation is two weeks.

Dyes	Initial conc.	initial ABC	Final ABC	Percentage of degradation	pH before	pH after
2,6-dichlorophenol-indophenol	6.1×10^{-5}	0.82	0.01	100%	7.8	6.3
Eriochrome black T	3×10^{-5}	1.8	0.01	100%	7.25	6.5
Methyl orange	3×10^{-7}	1.8	0.01	100%	8	7.4
Xylenol orange	3×10^{-5}	1.01	0.01	100%	10	9.4
Commercial dye	1%	0.65	0.65	0%	7	7

Table (3.1.10) Effect of sunlight on different organic compounds in the presence 0 .1 % H_2O_2 and time of degradation is three weeks.

Dyes	Initial conc.	Initial ABC	Final ABC	Percentage of degradation	pH before	pH after
2,6-dichlorophenol-indophenol	6.1×10^{-5}	0.82	0.01	100%	7.8	7.7
Eriochrome black T	3×10^{-5}	1.8	0.01	100%	7.25	7.1
Methyl orange	3×10^{-7}	1.8	0.01	100%	8	7.4
Xylenol orange	3×10^{-5}	1.01	0.01	100%	10	9.4
Commercial dye	1%	0.65	0.30	53%	7	6.9

Table (3.1.11)   Effect of   sunlight on different organic compounds in the presence 0 .1 % H_2O_2 and time of degradation is 30 days.

Dyes	Initial conc.	Initial ABC	Final ABC	Percentage of degradation	pH before	pH after
2,6-dichlorophenol-indophenol	6.1×10^{-5}	0.82	0.01	100%	7.8	6.1
Eriochrome black T	3×10^{-5}	1.8	0.01	100%	7.25	6.1
Methyl orange	3×10^{-7}	1.8	0.0	100%	8	7.4
Xylenol orange	3×10^{-5}	0.85	0.01	100%	10	9.4
Commercial dye	1%	0. 65	0.20	%69	7	6.8

Table (3.1.12) Effect of daylight on different organic compounds in the presence 0.1 % H₂O₂ and time of degradation is one week.

Dyes	Initial conc.	initial ABC	Final ABC	Percentage of degradation	pH before	pH after
2,6-dichlorophenol-indophenol	6.1×10^{-5}	0.82	0.01	100%	7.8	6.1
Eriochrome black T	3×10^{-5}	1.8	0.01	100%	7.25	6.1
Methyl orange	3×10^{-7}	1.8	0.01	100%	8	7.4
Xylenol orange	3×10^{-5}	0. 85	0.85	0%	10	10
Commercial dye	1%	0.65	0.65	0%	7	7

Discussion

Figs. (3.1.1), (3.1.2), (3.1.3), (3.1.4) and (3.1.5) displays decolorization of MO, EBT, ingredients water color E102-133, DID, and XO in the presence of hydrogen peroxide concentration = [0.1] M , at different time of irradiation by sunlight and the initial pH for these dyes respectively is 7.25, 7, 7.8, 7.3 and 10. It can be seen that the decolonization was rapid in case MO and EBT where the degradation after 8 hours for MO and after 10 hours for EBT but it would slow down in case XO, commercial dye, and DID. This is due to the hydrophilic characteristics of these dyes, and the negligible generation of hydroxyl radicals in the bulk liquid under the conditions in this study [S.K. Kansal, M.Singh, and D.Sud]. Therefore, in the absence of hydrogen peroxide, the degradation process is not allow. Tables explain pH for all dyes decrease after irradiation. After all irradiation the solutions for all dyes get colorless that mean dyes degradation to CO_2 and H_2O

The mechanism of reactions

Sunlight is available throughout the year and hence, it could be effectively photo-catalytic used for degradation of pollutants in wastewater. More over there is no material deterioration in case when sunlight is used as a radiation source [N.Daneshvar, D. Salari, and A.R. Khatee]. The detailed mechanism of dyes degradation reactive intermediate which is responsible for the degradation is hydroxyl radical ($OH^.$). It is either formed by the decomposition of H_2O_2 which is an extremely strong, non-selective oxidant (E = +3.06 V) which leads to the partial or complete mineralization of several organic chemicals [S.Sakthivel, B.Neppolian, B.V.Shankar and M. Palanichamy]

$$h^+_{VB} + H_2O_2 \rightarrow H^+ + OH^. \qquad (1)$$

$$OH^. + dye \rightarrow degradation\ of\ dye$$

3.2 Decoloration of Murexide by hydrogen peroxide

A- From Fig.(3.2.1) decoloration of the mixture is increased with the increasing the time. The maximum decoloration was about 100%. This is because hydrogen peroxide formed free hydroxyl radical ($OH^{0.}$), which make fast degradation of the dye. The detailed mechanism of the dye degradation is the formation of reactive intermediate, which is responsible for the degradation is hydroxyl radical (OH^0). It is either formed by the decomposition of H_2O_2 which is an extremely strong, non-selective oxidant (E = +3.06 V) and leads to the partial or complete mineralization of several organic chemicals [Sakthivel, 2003, 65]:

$$h^+{}_{VB} + H_2O_2 \rightarrow OH^0 + OH^0 \tag{1}$$

OH^0 + dye $\rightarrow$ degradation of dye

B- Effect of time of stirring on treatment Murexide by silica sand.

Fig. (3.2.2) Showed that in the initial stages, the decoloration rate is increased with increasing the time of stirring, decoloration of mixture is increased with the increasing of the time[Deng et al., 1998, 16 , Wu et al., 1998, 17; Zheng, et al., 2006, 58).

C- Effect of time of storing on treatment of Murexide with silica sand

Fig.(3.2.3) showed that in the initial stages, the decoloration increased with the increase of the time store[GB 11914-89, Omran, A.; El-Amrouni, A.O.; Suliman, L.K.; Pakir, A.H.; Ramli, M.; Aziz, H.A. Solid waste] .

D- Exposure time of stirring on treatment of Murexide with silica sand and activated carbon.

Fig (3.2.4) showed that in the initial stages, the decoloration is increased with the increase of the time. The process of adsorption is also influenced by the length of time that the AC is in contact with the contaminant in the wastewater. Increasing contact time allows greater amounts of contaminant to be removed from the water.

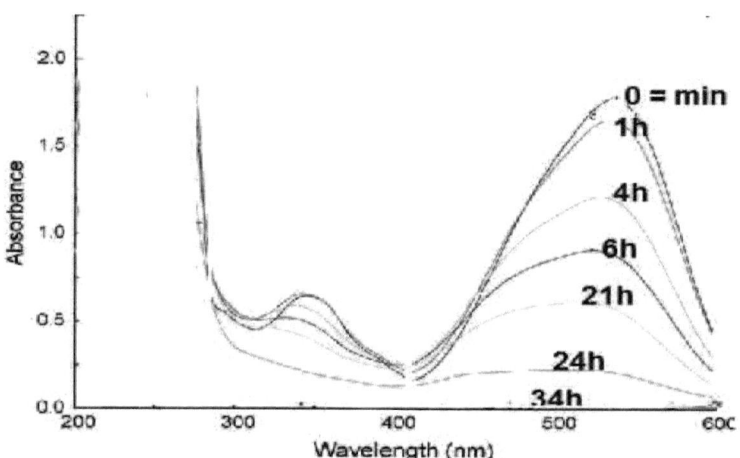

Fig.(3.2.1) Effect of time on decoloration of [Murexide] = 0.005M, [H₂O₂] = 30% at day light.

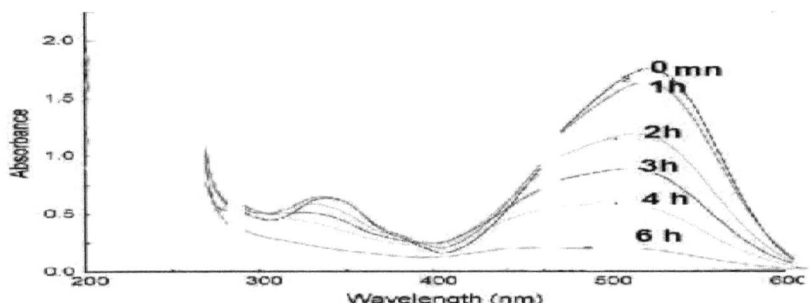

Fig.(3.2.2) Effect of time stirring on decoloration of [Murexide] = 0.005M, in present of silica sand.

38

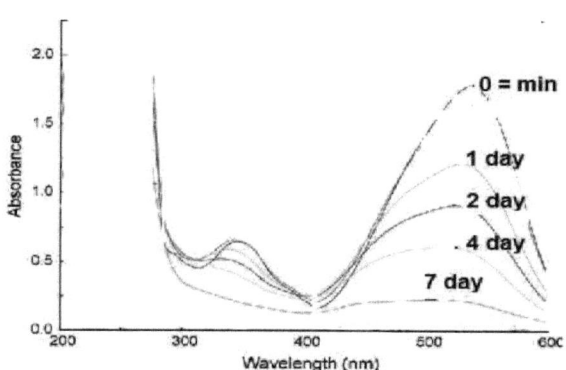

Fig.(3.2.3) **Effect of time store on decoloration of [Murexide] = 0.005M   in present silica sand at day light.**

39

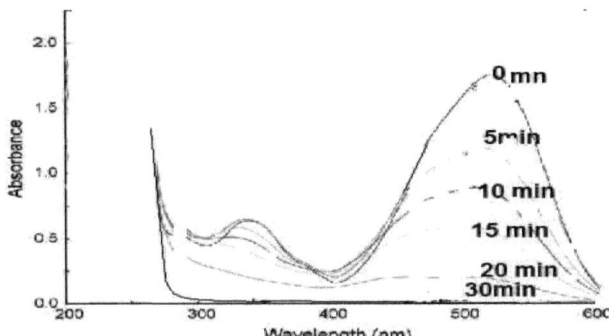

Fig.(3.2.4) Effect of time stirring on decoloration of [Murexide] = 0.005M, via silica sand and activated carbon at day light.

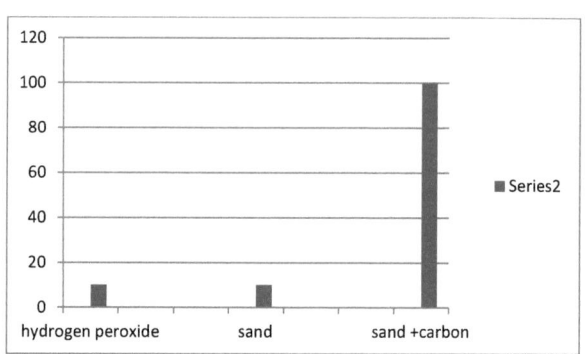

Fig(3.2.5). Effect of multi media of chemical compounds in removing Murexide (0.005M) after 30 min. of treatment.

The effect of activated carbon and sand give highest maximum decoloration was about 100% Fig.(5) explain that, the percentage of degradation was calculated from the following equation:

$$Degradation\% = [1-A_t/A_0] \times 100$$

Where A_0 initial absorbance and at A_t finial absorbance [Rashed et al., 2007, 73].

Activated carbon(AC) is extremely porous with a very large surface area. Certain contaminants accumulate on the surface of the AC in a process called adsorption. The two main reasons that chemicals adsorbed on the AC are a "dislike" of the water, and attracted to the AC. Many organic compounds, such as Murexide can be adsorbed by AC. AC also removed metals that are bound to organic molecules. It is important to note that carbon is not necessarily the same as AC. AC removes vastly more contaminants from water than does ordinary carbon[AL- Imarah et al., 2006, Omran, A.; El-Amrouni et al. *Eng. Manage. J.* 2009, *8*, 97-106.]

3.3 Effect of different adsorbents on removal of heavy metals from waste water

This effect was studied using of activated carbon, ceramic and silica (dose 1 g) at room temperature was shown in tables. The results show that the percentage of adsorption was increased when used a mixed of adsorbents. It is apparent that by increasing the number of sorption sites available for sorbent-bio – solute interaction is increased, there by resulting in the increased percentage of elements removal from the solution. But when used one adsorbent be that the percentage of adsorption was decreased and observed that there are high percentage of adsorption for some elements by one adsorbent like Hg^{+2} with activated carbon show table 1, pb^{+2} with ceramic show table 2 and Mg^{+2} with silica show table 3. When used three layers of adsorbents the percentage of adsorption was increased and obtained 100% removal for all elements ions show table 7. Table 8 displays that the best absorbent for Pb^{+2} is Ceramic, the best absorbent for Hg^{+2} is activated carbon , the best absorbent for Zn^{+2} and Ca^{+2} is silica [Al-Rawi, S.M. (1987)] and the best absorbent for Mg^{+2} is silica and activated carbon. Table 9 displays that the best absorbent for Pb^{+2} and Hg^{+2} is activated carbon with silica and ceramic with activated carbon[Caldorn, R.L and Moo, E.W (1992)], the best absorbent for Zn^{+2} and Mg^{+2} is silica with ceramic and activated carbon ceramic the best absorbent for Ca^{+2} is silica with activated carbon. Table 10 displays that the best filter for removal heavy metals is contain three layers of activated carbon with silica and ceramic[Chian, S. L (2002)] where obtain

Table (3.3.1) effect of Activated carbon on the percentage removal of elements and the amount of element adsorbed on adsorbent (qe).

element	Pb^{+2}	Hg^{+2}	Zn^{+2}	Ca^{+2}	Mg^{+2}
%Removal	70%	100%	80%	90%	90%
qe	0.035	0.05	0.04	0.045	0.045

Table (3.3.2) effect of ceramic on the percentage removal of elements and the amount of element adsorbed on adsorbent (qe).

element	Pb^{+2}	Hg^{+2}	Zn^{+2}	Ca^{+2}	Mg^{+2}
%Removal	100%	20%	64%	20%	80%
qe	0.05	0.01	0.032	0.01	0.04

Table (3.3.3) effect of silica on the percentage removal of elements and the amount of element adsorbed on adsorbent (qe).

element	Pb^{+2}	Hg^{+2}	Zn^{+2}	Ca^{+2}	Mg^{+2}
%Removal	90%	98%	90%	90%	98%
qe	0.045	0.049	0.045	0.045	0.049

Table (3.3.4) effect of Activated carbon with ceramic on the percentage removal of elements and the amount of element adsorbed on adsorbent (qe).

element	Pb^{+2}	Hg^{+2}	Zn^{+2}	Ca^{+2}	Mg^{+2}
%Removal	100%	100%	100%	56%	100%
qe	0.025	0.025	0.025	0.014	0.025

Table (3.3.5) Effect of Activated carbon with silica on the percentage removal of elements and the amount of element adsorbed on adsorbent (qe).

element	Pb^{+2}	Hg^{+2}	Zn^{+2}	Ca^{+2}	Mg^{+2}
%Removal	100%	100%	96%	98%	98%
qe	0.025	0.025	0.024	0.0245	0.0245

Table (3.3.6) Effect of ceramic with silica on the percentage removal of elements and the amount of element adsorbed on adsorbent (qe).

element	Pb^{+2}	Hg^{+2}	Zn^{+2}	Ca^{+2}	Mg^{+2}
%Removal	0.94%	80%	100%	80%	100%
qe	0.0235	0.02	0.025	0.02	0.025

Table (3.3.7) Effect of Activated carbon with silica and ceramic on the percentage removal of elements and the amount of element adsorbed on adsorbent (qe).

element	Pb^{+2}	Hg^{+2}	Zn^{+2}	Ca^{+2}	Mg^{+2}
%Removal	100%	100%	100%	100%	100%
qe	0.025	0.025	0.025	0.025	0.025

46

Table (3.3.8) Effect of different adsorbents on removal of heavy metals, Ca^{+2} and Mg^{+2}

elements	Initial concentratio n] molar	Conc. after Treatment by silica	Conc. after treatment by ceramic	Conc. after treatment by activated carbon
Pb^{+2}	0.01	0. 001	0	0.003
Hg^{+2}	0.01	0.0002	0.008	0
Zn^{+2}	0.01	0.001	0.0036	0.002
Ca^{+2}	0.01	0.001	0.008	0.001
Mg^{+2}	0.01	0.0002	0.002	0.001

Table (3.3.9) Effect of different two mixtures of adsorbents on removal of heavy metals, Ca^{+2} and Mg^{+2}

elements	Initial concentration	Conc. after Treatment by activated carbon and ceramic	Conc. after treatment by ceramic and silica	Conc. after treatment by activated carbon and silica
Pb^{+2}	0.01	0	0.0006	0
Hg^{+2}	0.01	0	0.002	0
Zn^{+2}	0.01	0	0	0.0004
Ca^{+2}	0.01	0.0044	0. 002	0.0002
Mg^{+2}	0.01	0	0	0.0002

Table (10) Effect of different mixture of three adsorbents on removal of heavy metals, Ca^{+2} and Mg^{+2}

elements	Initial concentration	Conc. after Treatment by activated carbon , ceramic and silica
Pb^{+2}	0.01	0
Hg^{+2}	0.01	0
Zn^{+2}	0.01	0
$Ca^{+?}$	0.01	0
Mg^{+2}	0.01	0

48

4. Conclusion

This is research focus on the development of a more reliable photo degradation that can be activated by solar light. In the present paper we have selected the UV photolysis of H_2O_2 as a source of hydroxyl radicals and investigated the decolonization of dye as a first step of a research amide at evaluating the applicability of this technique to the applicability of this technique to the treatment and re-use of dye house waste water. The degradation of various dyes and reports the main advances. H_2O_2 has been suggested to be efficient for the degradation of various toxic organic pollutants e.g dyes in water in the presence of sunlight. The findings also suggest that various operating parameters such as pollutant types and initial concentration, initial pH of the reaction medium, of dyes. In the present paper we have degradation 100% for all dyes at different times. The effect of activated carbon and sand give highest maximum decoloration was about 100% Fig. (3.2.5) explain that, the percentage of degradation was calculated from the following equation:

Degradation% = $[1-At/A^0] \times 100$

Where A^0 initial absorbance and at finial absorbance [Rashed et al., 2007].

Activated carbon (AC) is extremely porous with a very large surface area. Certain contaminants accumulate on the surface of the AC in a process called adsorption. The two main reasons that chemicals adsorbed on the AC are a "dislike" of the water, and attracted to the AC. Many organic compounds, such as Murexide can be adsorbed by AC. AC also removed metals that are bound to organic molecules. It is important to note that carbon is not necessarily the same as AC. AC removes vastly more contaminants from water than does ordinary carbon [AL- Imarah et al., 2006, Omran et al. 2009).

49

It can be concluded that it is necessary for various adsorbents to be tested because of their different surface properties in the determination of optimum conditions in terms of adsorbents for removal of the heavy metals by adsorption from aqueous solution without changing the conditions. Various adsorbents have different adsorption capacities for a few of the toxic metal ions. The best absorbent for Hg^{+2} is activated carbon is apparent that by increasing the number of sorption sites, pb^{+2} with ceramic and Mg^{+2} with silica. When used three layers of adsorbents the percentage of adsorption was increased and obtained 100% removal for all elements ions decrease in textural parameters like pore diameter and pore volume are known to be the influencing factors for the adsorption efficiency. These adsorbents materials are suitable for adsorption of toxic metal ions.

IV- References

References

AL- Imarah F. J., Sahil M. K. (2006), "Pollutions in the effect of Dairyand" Pollutants in the effluents of Dairy and soft drinks Industries in Basra city: there effect upon water of Shatt al Arab by column filled with sand and charcoal", Scientific research, SRO 4, Basra-Iraq.

Aziz, H.A.; Foul, A.A.; Isa, M.H.; Hung, Y-T. Physico-chemical treatment of anaerobic landfill leachate using activated carbon and zeolite Batch and column studies. Int. J. Environ. Waste Manage. **2010**, 5, 269-285.

Deng Nansheng, Wu Feng, Luo Fan *et al.* (1998), Photochemical oxidation and degradation of dyes by Fe(III)-hydroxy complexes, Environment and Exploitation, 12, 16-18 (In Chinese).

King W, Doddas L, Allen A (2000). Relation between stillbirth and specific chlorination byproducts in public water supplies. *Envi Health Pers*; 108(9): 67-78.

Marawski A W, Kalenezuk R (2000). Adsorption of trihalomethanes (THMs) onto carbon spheres. Desalination; 130: 107-112.

Niu Fusheng, Xu Xiaojun, Gao Jianguo, XU Jinqiu (2001), Study on purifying quartzite by mineral processing, Yunnan metallurgy, Vol.30(1), 18-21 (In Chinese).

Omran, A.; El-Amrouni, A.O.; Suliman, L.K.; Pakir, A.H.; Ramli, M.; Aziz, H.A. Solid waste management practices in Penang State: A review of current practices and the way forward.*Environ. Eng. Manage. J.* **2009**, *8*, 97-106.

Rashed, M. N., El-Amin, A. A. (2007), Photocatalytic degradation of methyl orange in aqueous TiO_2 under different solar irradiation sources, Int. J. Phys. Sci., 2(3), 73-81.

Sakthivel S., Neppolian B., Shankar M. V., Arabindoo B., Palanichamy M., Murugesan V. (2003), Solar photocatalytic degradation of azo dye: comparison of photocatalytic efficiency of ZnO and TiO_2, Solar Energy Materials and Solar Cells, 77(1), 65-82. Tala Suwa (2003), Used oxalic acid leaching and photocatalytic wastewater treatment to produce glass sand, Foreign metal mineral processing, 40, 40-43 (In Chinese).

USEPA,"Stage 1 Disinfectants and Disinfection by –products rule", office of water (EPA).Viessman, W. Hammer, M. Water supply and pollution control. Harper collies CollegePublishers. 15th edition. 1993: 485-488.

Weige Liao (2001), Used oxalic acid leaching to produce high-purity silica sands, Foreign metal mineral processing, 38, 33-36 (In Chinese).

Wietlik J, Stanis R, Bodzek M (2002). Adsorption of natural organic matter oxidized with ClO_2 on granular activated carbon. *Wat res*; 36(9): 2328-36.

Wu Feng, Deng Nansheng, Luo Fan et al. (1998),Study on discoloration of water soluble dyes induced by photolysis of Fe(III)-oxalate complexes, Environmental pollution contamination, 20(1), 17-20, (In Chinese).

Wu Huating, Xie Lishan, Xu Yafang, Chen Jiannan, Lin Gucheng, Zheng Xiaoan, Yan Guiyang (2010), Study on purification methods of removing impurity from quartz. Highlights of science paper online, 3(17), 1752-1756, (In Chinese).

Zheng Xiao-hong, Wang Yan, Wu Dang-lan, Chen Zhen, Lin Yu-man (2006), Studies on treatment of dyeing wastewater with polysilic aluminium ferric flocculants. Journal of Fujian Normal University (natural science), 22(4), 58-62, (In Chinese).

A.P. Pantelis, P.X. Nikolaos, M. Dionissios,Water Res. 40 (2006) 1276.

J. Ara~na, J.A. Herrera Melián, J.M. Do~na Rodríguez, O. González Díaz, A.

Viera, J. Pérez Pe~na, P.M. Marrero Sosa, V. Espino Jiménez, Catal. Today 76 (2002) 279.

H.D. Mansilla, C. Bravo, R. Ferreyra, M.I. Litter,W.F. Jardim, C. Lizama, J. Freer, J.Fernández, J. Photochem. Photobiol. A: Chem. 181 (2006) 188.

Pierce J., "Color in textile effluents - the origins of the problem", J. Soc. Dyers Color 110 (1994) 131.

Neamtu M., Siminiceanu I., Yediler A. and Kettrup A., "Kinetics of decolorization and mineralization of reactive azo dyes in aqueous solution by the UV/H_2O_2 oxidation", Dyes Pigm. 53 (2002) 93.

Sun J. H., Sun S. P., Wang G. L. and Qiao L. P., "Degradation of azo dye amido black 10B in aqueous solutions by Fenton oxidation process", Dyes Pigm. 74 (2007) 647.

Méndez-Paz D., Omil F. and Lema J. M., "Anaerobic treatment of azo dye Acid Orange 7 under fed-batch and continuous conditions", Water Res. 39 (2005) 771.

Brown M. A. and DeVito S. C., "Predicting azo dye toxicity", Critical Reviews in Environmental Science and Technology 23 (1993) 249.

Selvam P. P., Preethi S., Basakaralingam P., Thinakaran N., Sivasamy A. and Sivanesan S., "Removal of rhodamine B from aqueous solution by adsorption onto sodium montmorillonite", J. Hazard. Mater. 155 (2008) 39.

Jae-Wook L., Seung-Phil C., Ramesh T.,Wang-Geun S. and Hee M.,"Evaluation of the performance of adsorption and coagulation processes for the maximum removal of reactive dyes" Dyes and Pigments, 69 (2006) 196.

Pagga U. and Drown D., "The degradation of dye-stuffs. Part II. behaviour of dyestuffs in aerobic biodegradation tests", Chemosphere 15 (1986) 479.

Namboodri C. G., and Walsh W. K., "Ultraviolet light/hydrogen peroxide system for decolorizing spent reactive dye bath waste water", American Dyestuff Reporter (1996).

Fenton H., "Oxidation of tartaric acid in presence of iron", J. Chem. Soc. Trans., 65 (1894) 899.

I.A. Alaton, I.A. Balcioglu and D.W. Bahnemann .Advanced oxidation of a reactive dyebath effluent: comparison of O_3 H_2O_2/UV-C and TiO_2/UV-A processes.J. Water res. 36: 1143-1154, (2002).

M. Muruganandham and M. Swaminathan Photochemical oxidation of reactive azo dye with UV/H_2O_2 Oxidation. Dyes and Pigments 62: 269.(2007).

S.K. Chaudhur and B. Sur Oxidative decolonization of reactive dye solution using fly ash catalyst. J. Environ. Engg. 126: 583,(2000).

D. Georgiou, P.Melidis and A. Aivasidis. imouhopoulosm,Degradation of azo –reactive dyes by UV in the presence of H_2O_2 J. Dyes and Pigments 52: 69,(2002).

Y. Yang, D.T.Wyatt and M. Bahorsky ecolonization of dyes using UV/H₂O₂ photochemical oxidation. J. Textile Chem. Colorist 30: 27,(1998).

B. Neppolian, H.C. Choi and S. Sakthivel V. Murugesan, Solar/UV- induced photocatalytic degradation of three commercial textile dyes. J. Hazard. Mater. B 89: 303,(2002).

Y. Wan Solar photocatalytic degradation of eight commercial dyes in TiO₂ suspension. J. Water res. 34 : 990,(2000).

M. Neamtu, I. Siminiceanu and A. Yediler Kinetics of ecolonization and mineralization of reactive azo dyes in aqueous solution by the UV/H₂O₂ oxidation, J. Dyes and Pigments, 53: 93-99, (2002).

H.Nilsum Critical effect of hydrogen peroxide in photo chemical dye degradation J. Dyes Pigments 41:225-230,(1999).

S.Yana, Wangp, yangx, Shanl, and Zhanndgw Degradation efficiencies of azo dye acid orang 7 by the interaction of heat , UV and anions with common oxidants: per sulfate proxy mono sulfate, and hydrogen peroxide J. Hazard Mater , 179: 1-3, (2010).

G. M. Colonna and T. Caronna. Oxidadative degradation of methyl Orange dye by UV in the of hydrogen peroxide. J. Dyes and Pigments 41, 211-220,(1999).

C. Jeed, and A. Mahata Demineralization of organic pollutants on he dye modified TiO_2 semiconductor particulate system using visible light. Appl.Catal._benviron 33(2) :119-125 ,2001

M.N Rashed and A.A. El-Amin.Photocatalytic degradation of methyl orange in aqueous TiO_2 under different solar irradiation sources. J. physical Sciences. 2: 73-81 (2007).

S.K. Kansal, M.Singh, and D.Sud. Studies on photodegeradation of two commercial dyes in aqueous phase using different photocatalysts. J. Engineering and technology 10:1016, (2006).

N.Daneshvar, D. Salari, and A.R. Khatee Photocatalytic degradation of azo dye acid red 14 in water on ZnO as an alternative catalyst to TiO_2 J. Photobio.Chem.157 : 111, (2003).

S.Sakthivel, B.Neppolian, B.V.Shankar and M. Palanichamy solar photocataytic degradation of azo dye: comparison of photocatalyiz efficiency of ZnO and TiO_2. J. Sol Ener Mater, Sol. 77:68,(2003).

Al-Rawi, S.M. (1987) "turbidity removal of drinking water by dual media filtration" thesis submitted to civil engineering Dep,

Caldorn, R.L and Moo, E.W (1992) " Activated carbon Filtration Equipment", North dakoa University, U.S, PP.701-788.

Chian, S. L (2002)"appropriate microbial indicator test for drinking water in developing countries and assessment for ceramic water filter" thesis submitted to Environmental Engineering department, PP. 291- 297.

Printed by Books on Demand GmbH, Norderstedt / Germany